ISBN 978-1-5279-2523-6
PIBN 10899114

Forest Service

Intermountain Forest and Range Experiment Station

Research Paper INT-278

August 1981

Correlating Laboratory Air Drop Data with Retardant Rheological Properties

Wayne P. Van Meter
Charles W. George

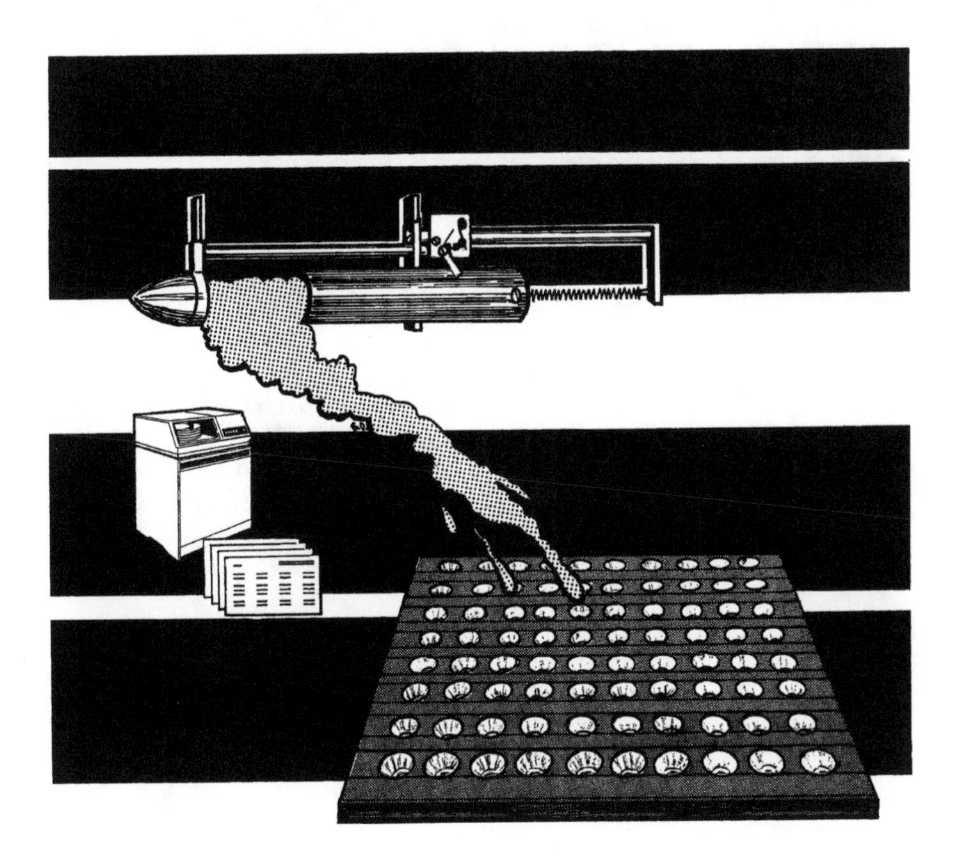

Research Summary

Based on the hypothesis that the spatial distribution on the ground of fire retardant materials, dropped from fixed-wing aircraft, must be a result of the physical properties of the retardant, a series of experiments has been run to measure the dispersal patterns obtained with materials of known density, viscosity, yield stress, surface tension, effective viscosity, and modulus of elasticity. The objective is the capability of predicting the relative size of the dispersal pattern from measurements made in the laboratory.

The experiments employed a 40 mi/h (64.4 km/h) wind tunnel airstream and 100 ml samples released 30 inches (76.2 cm) above the surface. An array of cups recessed into the surface trapped samples for weighing.

Correlation of the data shows that each of the six physical-chemical properties mentioned contributes to determining the size of the dispersal pattern. Deleting any one of them degrades the quality of the correlation, but effective viscosity (or elasticity) and density seem to be more important than the others. From other work (Andersen and others 1976) the effective viscosity has been found important in determining the droplet size distribution resulting from dispersal at any given aircraft speed.

Recommendations are made for future study that include refined wind tunnel experiments and correlation of these with accurately monitored full-scale field tests.

The Authors

WAYNE P. VAN METER has been a faculty member of the University of Montana, Department of Chemistry, since 1959. He also conducts occasional summer research studies for the Fire Control Technology research work unit at the Northern Forest Fire Laboratory in Missoula. He received his B. S. and M. S. degrees from Oregon State University and, in 1959, his Ph.D. degree in chemistry from the University of Washington.

CHARLES W. GEORGE graduated from the University of Montana in 1964 in forest engineering. He received his M. S. degree from the University of Montana in 1969. In 1965, he joined the Intermountain Station's Northern Forest Fire Laboratory in Missoula, Mont., where he has conducted studies related to prescribed fire, pyrolysis and combustion, fire retardant chemicals, and aerial fire retardant delivery systems. He is currently the leader of the Fire Control Technology research work unit.

Contents

United States
Department of
Agriculture

Forest Service

**Intermountain
Forest and Range
Experiment Station**

Research Paper
INT-278

August 1981

Correlating Laboratory Air Drop Data with Retardant Rheological Properties

Wayne P. Van Meter
Charles W. George

INTRODUCTION

Aircraft of several types have been used for nearly 30 years to deliver water or water-based mixtures onto fuel materials in the course of attempts to control or extinguish wildfires. Over the years, numerous revisions have been made of techniques and changes of mixture composition. Most resulted from attempts to improve the distribution of the retardant, its wetting or coating action on the fuel, or its actual performance once it had been applied.

For example, it was recognized early that the thickness, or viscosity, of the retardant has some effect on the behavior of a freefalling, coherent mass of liquid (Davis 1959). Full-scale field tests in 1955 to 1959, using simultaneous drops of different compositions, showed by simple comparison that thickened materials reached the ground in a more coherent mass and in a shorter total length of time than water alone.

Other studies (MacPhearson 1967, Swanson and others 1978) have been directed toward characterizing aircraft and tank mechanism performance. There have been several experimental studies (Wilcox and others 1961) and at least one theoretical study (Garcia and Wilcox 1961) of the behavior of liquid volumes of various sizes falling from rest in still air.

In actual practice, viscosity is the most common, and usually the only measurement used to monitor and control the mix of retardant formulations in field operations. Although viscosity can usually be related to a formulations preparation, it is known that the viscosity of a retardant does not directly relate to its aerial delivery characteristics and performance. It may be possible, however, through quantification of additional retardant physical-chemical (rheological) properties, and using new and more sophisticated instrumentation and techniques, to determine and relate more appropriately these properties to the retardants full-scale field performance.

The purpose of the work has been to explore some measurements that could be made and to identify those that more appropriately relate to aerial delivery, deformation, breakup, dispersion, and distribution on the ground under actual drop conditions. Most of the experimental work was done in 1970 and reported in an in-house document by the Northern Forest Fire Laboratory. Since then, instrumentation has become available for measuring the elastic properties of fluids, and other important related studies have been conducted (Andersen and others 1974a, 1974b, and 1976) placing an additional value on the original effort. The present report embodies recent (1979) elasticity data, considers the findings of Andersen and others (1974a, 1974b, and 1976) and includes a more thorough data analysis. In the following two sections, the equipment and material used will be described and their performance qualitatively, but critically, described.

EQUIPMENT AND MATERIALS

Attapulgite clay-water, Colloid 26 (modified polysaccharide)-water, and CMC (sodium carboxymethylcellulose)-water mixtures were prepared from material samples supplied by manufacturers from stocks used in blending brand name retardant products. The Fire-Trol™ retardant sample came from stocks at the Missoula Region 1 Airtanker Retardant Base. The Arcadian Poly-N™ used was from a test sample of 10-34-0 concentrate supplied by the manufacturer. The Phos-Chek™ samples also were supplied directly by the manufacturer. The CMC type is a standard product and the Colloid 26 type comprises three samples containing enough thickener to produce Brookfield viscosities of about 800, 1500, and 2200 centipoises.[1] In each case, the formulation was made by adding the indicated amount (table 1) a little at a time for about 1/2 minute into 500 ml of distilled water in a Waring blender running at its "slow" setting for exactly 2 minutes. The mixture was stored overnight in a closed jar before being used in wind tunnel tests or in physical property measurements.

[1]Brookfield viscosity values were obtained from a portable Brookfield Viscometer model LVF equipped with spindle 4 and allowed 1 minute of rotation at 60 r/min before making the measurement.

Table 1.--Test samples and their mixing ratios

Sample	Material	Amount/ 500 ml water
Att 1	Attapulgite clay	80.0 g
Att 2	Attapulgite clay	40.0 g
Att 3	Attapulgite clay	25.0 g
Att 4	Attapulgite clay	60.0 g
FT 1	Fire-Trol 100	166.0 g
FT 2	Fire-Trol 100	166.0 g
CMC 1	CMC	7.0 g
CMC 2	CMC	3.5 g
CMC 3	CMC	5.0 g
CMC 4	CMC	2.0 g
PC 7	Phos-Chek 202	68.3 g
PC 3	Phos-Chek 202	68.3 g
XA 7	Colloid 26	8.0 g
XA 3	Colloid 26	2.0 g
PC 3	Phos-Chek 202XA	68.3 g
PC 4	Phos-Chek 202XA	68.3 g
PC 5	Phos-Chek 202XA	68.3 g
PN 1	10-34-0 concentrate (Arcadian Poly-N)	167 ml
PN 2	10-34-0 concentrate (Arcadian Poly-N)	84 ml
PN 3	10-34-0 concentrate (Arcadian Poly-N)	125 ml

Note: Phos-Chek XA formulations were thickened with Colloid 26.

Densities were measured in Pyrex pycnometers containing about 30 ml each.

Using a Haake Rotovisco™, rotational viscometer, data could be obtained from which the yield stress could be derived. By varying the speed of rotation of the rotor, values of shear stress were observed for a series of values of the rate of shear. When these values are plotted and the curve extended to zero rate of shear the shear stress axis intercept is defined as the yield stress. An alternative method, which yields similar values, involves plotting on log-log coordinates the shear stress against the rate of shear and finding the stress at the point at which the curve departs from linearity. A medium viscosity measuring beaker and MV-I and MV-II rotors were used.

In order to measure the surface tension of the liquids, a tensiometer of the "jolly balance" type was devised (fig. 1).

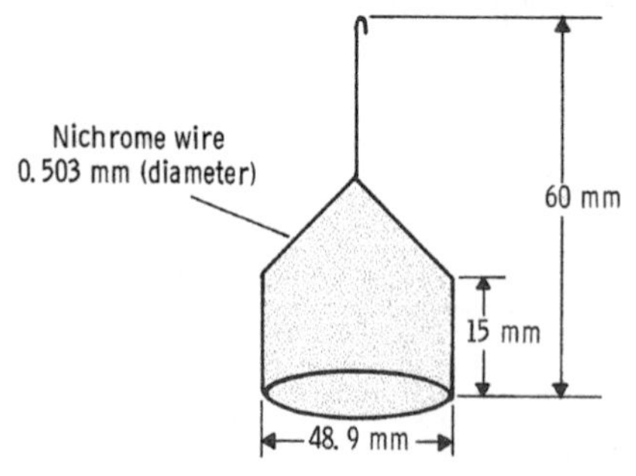

Figure 1.--Tensiometer ring.

The sample vessel, a 4-in (10 cm) Petri dish, rests on a platform about 8 in (20 cm) long, hinged to a solid support on one end and suspended by a waxed linen cord at the other end. The cord is fastened to the shaft of a synchronous motor. The motor speed and shaft diameter are chosen so that the vertical (downward) rate of motion of the center of the sample is about 12 mm per minute. A wire ring is suspended in such a way that the plane of the ring is parallel to the surface of the sample and its weight is borne by a microbalance-arm attachment on a Statham strain gage. The gage has a 60-g capacity, while the 2-3/4-in (7 cm) beam arm has a mechanical advantage of 10.

The strain gage signal is amplified by a Statham Universal Readout (Model UR5) and recorded by a Bausch and Lombe VOM5 recorder. Calibration is effected by adjusting sensitivity and balance controls until hanging a 3-g weight on the balance hook (the ring already being in place) causes the recorder pen to move from exactly zero to exactly full scale.

Air velocity in the wind tunnel was controlled at 40 mi/h (64.4 km/h) (± 0.5 mi/h (± 0.8 km/h) estimated fluctuation limit). The conditions of temperature, pressure, and humidity that were not controlled were in the ranges of 82° to 85° F, 680 to 685 mm Hg (total), and 20 to 22 percent relative humidity. The wind tunnel is of square cross section, 36 by 36 inches (91.4 cm).

The sample release mechanism holds 98.4 ml of liquid. It consists of a spring-actuated cylindrical container which, on release, moves away from a flat, gasketed end plate, sliding over a fixed piston. The chamber is 3.7 in (94 mm) in length by 1.4 in (36.5 mm) inside diameter. The liquid contained in the horizontal cylinder and between the end plate and the piston is thus left free to fall with zero initial momentum. Liquid-tight closure is provided by a 1/16- by 1-3/8 in (1.6 by 34.9 mm) rubber O-ring recessed into the edge of the piston and by a silicone rubber facing (G.E. "RTV") on the end plate. The mechanism is filled through a 1/2-inch (12.7 mm) hole in the end plate. Figure 2 shows the sample release mechanism in a closed and open or release position.

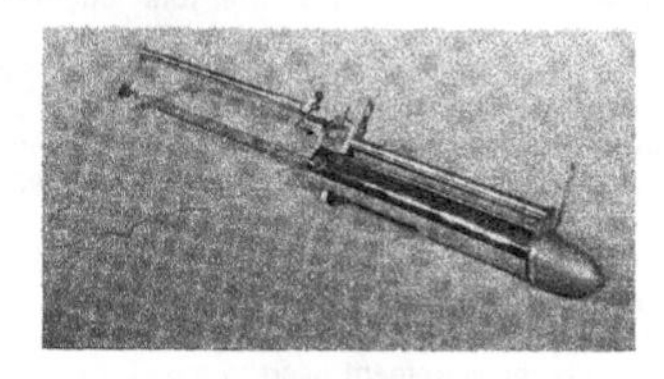

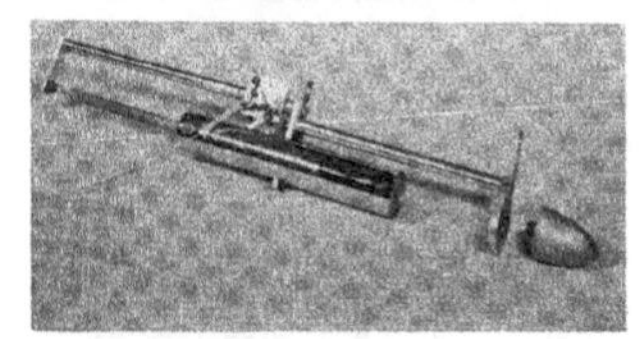

Figure 2.--Sample release mechanism shown in a closed (a) and release (b) position.

The pattern measurement system is an array of 480 poly-lined paper cups (average lip diameter 4.59 cm) positioned in 1-3/4 in (4.44 cm) holes spaced 3 in (7.62 cm) center-to-center in 1/4-in (0.635 cm) plywood sheets (fig. 3). The array is 10 cups wide and 48 long (parallel to wind direction). The plywood is mounted on 1-in (2.54 cm) blocks to allow the cup rims to be flush with the upper surface of the tray. Masking tape strips are effective in preventing dislodging of cups by holding the upwind edge of the lip against the tray. The vertical distance between the center line of the release mechanism cylinder and the tray surface is 30-1/2 in (77.5 cm).

Figure 3.--Sample cup tray, one of six trays that are placed side by side and used simultaneously.

Photographic observation of the release of liquid in the wind tunnel was conducted using two different camera and light combinations. In order to determine whether the release mechanism functioned rapidly enough and to observe directly the sequence of motions and shapes during dispersal, a high speed motion picture camera was used (Traid 200). Four 320-watt floodlights were located inside the tunnel, downwind, about 6 ft (1.83 m) from the release point. Another light (500 watts) was located outside the tunnel, facing inward at the lower left corner of the tunnel window, making an angle of about 60 degrees with the camera's line-of-sight. Camera-to-subject distance was about 10 ft (3.05 m). The film was Kodak High Speed Ektachrome Type B. Four different materials (CMC 2, CMC 4, PC 1, and water; see compositions and properties in table 1 and table 2) were each photographed twice, once with a 25-mm lens and once with a 75-mm lens.

To make measurements of the relative distributions of droplet sizes in midair after breakup of the main liquid mass, a Graflex Crown Graphic camera with a Kodak Ektar f4.5 lens was employed at f22. Subject distance was about 3-1/2 ft (1.07 m). Stroboscopic light sources (Synctron 200B, Dormitizer Co.) were used, one to the left at about 45 degrees to the camera direction, the other below and slightly to the right.

To avoid a highlight reflecting from the back wall of the tunnel, the camera was aimed upward at a slight angle (15 or 20 degrees from horizontal). Film was Polaroid Land Packets Type 55 P/N.

The apparent sizes of droplets, as viewed through the eyepiece reticle of a 12.5x binocular microscope, were compared to that of a 12-in (30.5 cm) ruler, photographed suspended under the release mechanism.

Equipment Performance

The measurement of viscosity with the Brookfield instrument has been, and should remain, popular because it is rapid, simple, and reproducible. Since the retardant materials are non-Newtonian in behavior, the viscosity as measured with the Rotovisco can agree with that of the Brookfield at only one rate of shear (for the one chosen Brookfield shear rate).

The use of the Rotovisco to measure yield stress is briefly mentioned by Van Wazer and others (1963). According to W. W. Morgenthaler (personal conversation), this method involves rapid engaging and disengaging of the rotor drive gears, producing enough rotation of the drive shaft to give a dynamometer signal (measurement of torque), but not enough to move the rotor much at all. The yield stress, then, is the maximum force that can be applied to the rotor without causing rotation. In the work being reported here, attempts also were made to measure yield stress by slowly turning the beaker, by hand, and observing the stress produced before any shearing occurred. Neither of these techniques worked well. It is not easy to rotate the shaft a desired amount with the gear lever, and duplicate run data did not agree. The log-log graphs of shear stress against shear rate are smooth enough to be believable, it is almost always possible to identify the point where linearity ceases and duplicates give similar values. Samples that contain large amounts of attapulgite clay yield more uncertain results because shearing by the measuring instrument increases viscosity and yield stress.

The measurement of surface tension was verified by comparing experimental results on certain pure liquids with values found in standard reference books. For example, the accepted value for water at 25° C is 72.3 dynes/cm. The observed value was 73.2 dynes/cm. Similar comparisons were seen for ethanol, chloroform, and acetone. The apparent value was independent of the rate of motion of the sample platform, both for rates somewhat slower and considerably (two or three times) faster than the one used here. The effect of temperature on the surface tension is only a few hundredths of a dyne per centimeter over a range of several degrees. This is at least 10 times smaller than usual differences between replicate samples. It is quite important, however, to be sure that the sample vessel and the ring are carefully washed, thoroughly rinsed, and untouched prior to each run. It is also true that careful attention must be paid to preserving the shape of the wire ring and its horizontal attitude when suspended from the balance hook. This adjustment also includes achieving, as nearly as possible, simultaneous release of all parts of the ring from the liquid surface. Accuracy is lost if the ring breaks loose from one side and swings.

To secure dependable operation of the mechanism used to release the retardant sample in the wind tunnel airstream, it was necessary to lubricate the O-ring with a few drops of light mineral oil, which was applied to the inside of the cylinder behind the piston. It is likely that the need for lubricant could be eliminated by using an O-ring with slightly larger diameter and by adjusting the groove depth to cause slightly less pressure against the inner wall of the cylinder.

Table 2.--Sample composition, properties, and dispersal pattern areas

Sample			Date	Composition	Density (*d*)	Brookfield viscosity (N_b)	Yield stress (*Y*)	Surface tension (*S*)	Dispersal pattern areas D>10	D>1.8	Total, D>T
					g/ml	*Centipoise*	*Dyne/cm²*	*Dyne/cm*	-------*Percentages*-------		
	Att	1A	7/16	13.8 percent clay	1.071	4240	450	73.0	2.3	9.4	37.7
	Att	1B	7/29	13.8 percent clay	1.075	[1]N.M.	480	77.2	2.5	9.4	38.8
Attapulgite	Att	2A	7/15	7.41 percent clay	[2]1.041	490	37	71.6	2.9	11.7	58.8
+ water	Att	2B	7/29	7.41 percent clay	1.039	670	72	73.4	4.6	22.7	55.4
	Att	3A	8/1	4.76 percent clay	1.025	190	5	71.9	3.5	18.1	65.4
	Att	4A	8/22	10.7 percent clay	1.046	2865	320	75.1	N.M.	N.M.	N.M.
Fire-Trol	FT	1A	7/29	9.0 percent clay	1.144	2240	170	78.3	5.6	36.5	76.5
100	FT	1B	8/1	9.0 percent clay	1.140	2415	200	76.7	4.6	29.2	80.4
	FT	2A	8/22	9.0 percent clay	1.147	2850	230	77.8	N.M.	N.M.	N.M.
	CMC	1A	7/16	1.38 percent CMC	[2]1.002	780	20	70.5	4.4	11.1	36.5
	CMC	1B	7/29	1.38 percent CMC	1.004	1005	20	72.7	4.6	18.8	35.6
	CMC	2A	7/15	.70 percent CMC	[2].999	150	20	72.4	4.2	15.8	37.9
CMC + water	CMC	2B	7/24	.70 percent CMC	.999	155	20	72.3	1.9	22.1	69.4
	CMC	3A	8/22	.99 percent CMC	1.003	260	10	73.1	N.M.	N.M.	N.M.
	CMC	4A	7/24	.40 percent CMC	[2].999	55	30	72.9	4.0	14.6	63.6
Phos-Chek	PC	1A	7/24	1.02 percent CMC	1.070	410	5	64.6	3.8	10.2	76.5
202 (CMC)	PC	1B	7/30	1.02 percent CMC	1.070	650	20	63.2	4.6	11.7	69.2
	PC	6A	8/22	1.02 percent CMC	1.075	1125	60	70.9	N.M.	N.M.	N.M.
Colloid 26	XA	1A	7/30	1.57 percent XA	[2].987	9050	800	N.M.	4.0	5.8	10.0
	XA	3A	8/1	.40 percent XA	[2].999	200	15	71.1	5.0	27.7	61.7
	PC	3A	7/30	[3]800 cP	1.059	N.M.	50	70.1	4.8	20.6	63.1
	PC	3B	7/31	[3]800 cP	1.070	1040	70	71.4	5.4	26.1	66.1
Phos-Chek	PC	4A	7/31	[3]800 cP	1.070	1680	120	69.9	5.6	14.6	59.4
202 XA	PC	4B	7/31	1500 cP	1.071	845	100	70.3	4.2	17.7	66.3
(Colloid 26)	PC	4C	8/22	1500 cP	1.068	1555	75	67.4	N.M.	N.M.	N.M.
	PC	5A	7/31	[3]2200 cP	1.070	2010	180	71.6	6.7	26.5	60.4
	PN	1A	7/23	1:3	1.119	25	8	76.6	6.3	36.7	74.0
Arcadian	PN	1B	7/25	1:3	1.118	25	8	79.3	6.3	29.0	66.7
Poly-N	PN	2A	7/23	1:6	1.069	25	8	79.0	1.9	19.8	64.8
10-34-0	PN	2B	7/25	1:6	1.081	25	8	79.0	1.9	19.8	64.8
	PN	3A	8/22	1:4	1.103	25	8	77.6	N.M.	N.M.	N.M.
	W	1A	7/22		.999	0	0	73.2	1.3	8.8	59.2
	W	1B	8/1		.999	0	0	73.2	3.5	30.6	67.3
Water	W	2A	7/22		.999	0	0	73.2	2.9	18.8	63.8
	W	2B	7/30		.999	0	0	73.2	3.3	18.1	63.6

[1]N.M. Values not measured or omitted due to equipment malfunction.
[2]The indicated data are used for further analysis which is discussed later.
[3]Special mixtures supplied by the manufacturer. Thickener has been adjusted to give the specified viscosity.

Because of the adherence of some of the sample material to the end plate and piston of the release mechanism and the spreading of some along the outside of the cylinder by the airstream, the volume released into the air is about 90 percent of the cylinder's volume. This applies to Phos-Chek and Fire-Trol of nominal viscosities.

The average length of time for complete opening of the sample release mechanism was 0.082 second (as measured from high-speed photographs of several retardant releases). The release was fast enough that the motion of the main portion of the sample was negligible during movement of the cylinder. The front portion of the sample, within a centimeter or so of the end plate, had begun to fall and to enter the airstream by the time the cylinder reached the end of its travel.

The impact of the cylinder on the recoil pads causes the whole mechanism to vibrate. No disturbance occurs at the front, because the sample has moved out of contact with the end plate. At the rear, a small wave or ridge of liquid is sometimes propelled downward by the piston moving against the portion of the sample adjacent to it. Sample cups were handled and weighed as quickly as possible to minimize error caused by evaporation of water. A few double weighings of cups, containing small, but measurable, amounts of material, showed that losses ranged from 10 to 20 percent. Losses from larger samples were much

There are three major types of ingredients in one column and three measurable physical properties in the other. Viscosity and yield stress seem always to be related, although not always directly proportional to each other. All of the thickeners have an effect on viscosity and yield stress; none have much effect on the surface tension of the mixture. The polymers and gums have essentially no effect on density, whereas the clays have considerable effect. The concentration of salts affects density and surface tension, but has virtually no influence on viscosity. Any surface active agent has a more pronounced effect on surface tension than either of the other two ingredient types has on any of the measured properties.

Distinct differences can be observed among the several materials tested, both in the sequence of events in midair during breakup and in the dispersal pattern on the impact surface. These differences can only be due to mechanical effects (forces) acting between the two fluids as the retardant penetrates the airstream. Thus, it is reasonable to expect variations in physical properties to be associated with changes in dispersal behavior. The simplest of such relationships would be a linear or regular curvilinear trace when numerical values of two such quantities are graphically compared.

Yield Stress and Viscosity Versus Area

In figures 4 and 5, the variations of the total area covered with yield stress and viscosity are shown. The data for Colloid 26 are meager, and the general shape of the curve is inferred by those of the other materials. The trends are in agreement with experience and intuitive expectations; the more resistant a fluid is to shearing forces, the smaller will be the volume of space occupied by droplets when the initial breakup and dispersal has used up all the momentum imparted by the aircraft's forward speed.

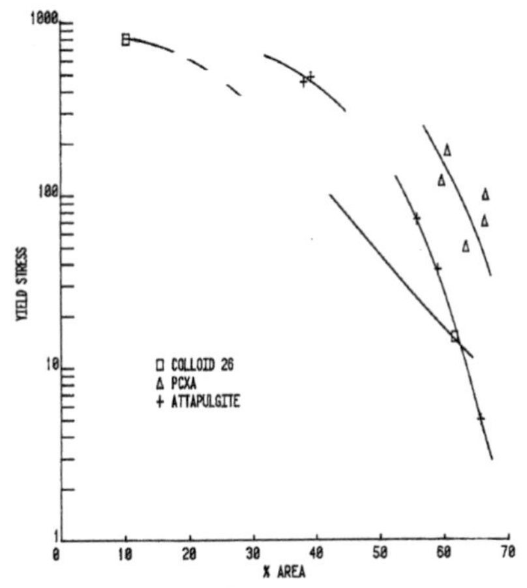

Figure 4.--Pattern coverage as a function of yield stress.

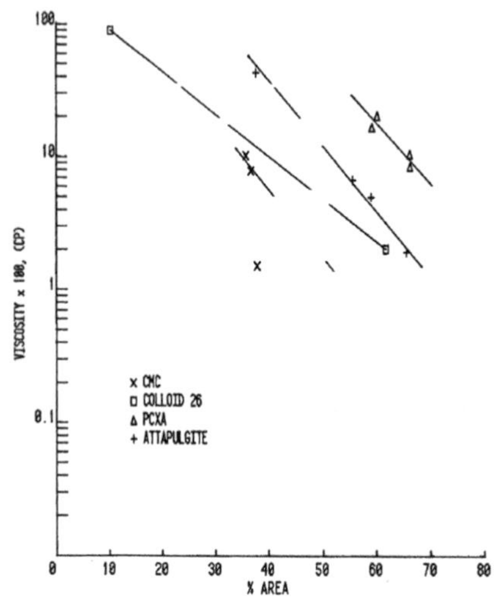

Figure 5.--Pattern coverage as a function of viscosity.

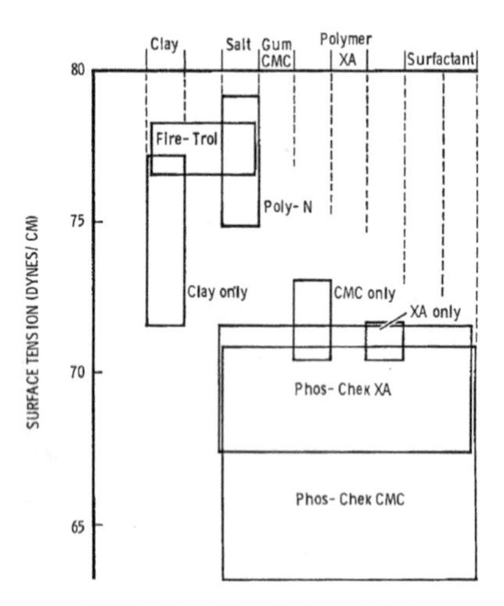

Figure 6.--Surface tension of test samples. Rectangles indicate the primary ingredients and the ranges of surface tension values observed.

Surface Tension

When all the samples run were considered in the order of their surface tensions, a significant grouping appears (fig. 6). The precision of the measurement is felt to be in the order of ± 2 dyne/cm. Nearly all of the values for the two-component mixtures (water and CMC, attapulgite clay, or Colloid 26) lie within ± 2 dynes/cm of the value for water alone (73.2 dynes/cm). When ammonium phosphate alone is present (as in 10-34-0) the surface tension is high. When clay is present with the salt (Fire-Trol), the salt has the same effect it has when alone. The marked depression of the surface tension in Phos-Chek mixtures cannot be logically ascribed to CMC and Colloid 26; neither has any effect alone. The corrosion inhibitor in all Phos-Chek formulations has a chemical identity that should cause it to have considerable potency as a surface active agent. When this material is solublized in distilled water at the same mass per volume concentration at which it occurs in Phos-Chek retardant mixed for use, the observed surface tension is about 60 dynes/cm.

Droplet Size Distribution

The measurement of the size distribution of droplets during the breakup process produced the graphs shown in figure 7. A high, narrow peak would indicate that the material tended to produce a uniform spray and that many drops would be of about the same size. The highest frequency observed was for 1.4-mm drops of pure water. Two other materials showed values nearly as high, however, and the difference may not be significant. It is clear that a trend exists, with high viscosity in the absence of salt (Att) producing the widest range of sizable frequencies. As salt content is increased, or as viscosity is decreased, the shape and position of the drop size distribution curve approaches that of water.

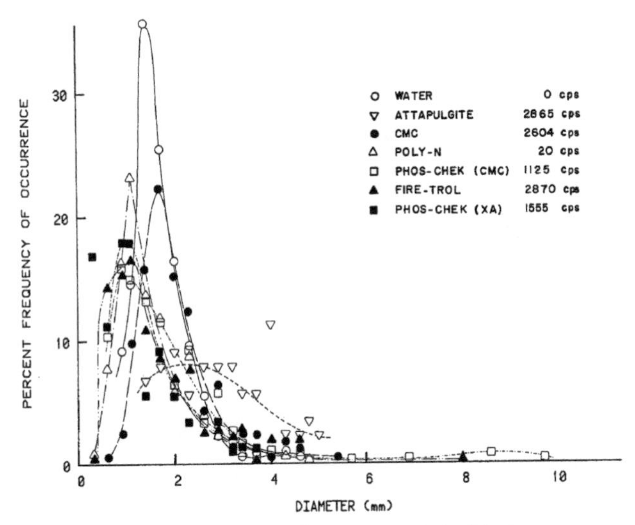

Figure 7.--Frequency of occurrence of droplets by diameter.

The droplet size distribution results should be regarded with caution for two major reasons. The photographic-optical method of comparison was not verified by any independent means to ensure that parallax or image quality was not introducing sizable error. Also, the object-camera distance was short enough that the limited depth of field caused only the drops near the centerline of the pattern to be in focus. Many were not visible at all, and size perception may have been incorrect for those not in sharp focus that were measured. In addition, a total of only 1,450 drops were measured from seven photographs. The statistical base is not strong, and reproducibility was not demonstrated.

Pattern Area Predictions

Correlation of the density, surface tension, and rheological properties with the drop pattern areas was accomplished by a computer program which generated the best linear least squares fit and supplied an analysis of variance and the coefficients of the equations. The data were handled in two phases. The first utilized 21 sets of data (those in table 2, excluding those for which pattern areas are not given, Poly-N, and water). The expression used in the computations has the form:

Dispersal pattern area = D (density) + N_b *(Brookfield viscosity)* + Y (yield stress) + S(surface tension) + constant, where the symbols D, N_b etc. are coefficients determined as a result of the analysis, and the parenthesized quantities identify numerical data items.

The results are given in the first three lines of table 3, where the main body of the table displays the coefficients of the above equation which yield the regression coefficient, r^2, found in the last column.

The second phase employed elasticity data. Samples were prepared (during 1979), using the original stocks of clay, CMC, and Colloid 26, that replicated the mixtures that produced some of the pattern area data in table 2. Elasticity and viscosity data were measured on these 1979 samples. It is assumed that values of the effective viscosity and of the modulus of elasticity, calculated from elasticity measurements made in 1979, can be used with the pattern areas and other data measured in 1970 on essentially identical materials. Six of the samples shown in table 2 have been chosen as those most closely resembling the 1979 samples in terms of their Brookfield viscosities, yield stresses, and surface tensions. They were identified in table 2 (see footnote 2) and by sample number in Part A of table 4. The values given for the effective viscosity and the modulus of elasticity were determined from original Rotovisco data using several known relationships:

Apparent viscosity $(N_a) = U\,S_c\,k$

where:
- U = the rotational speed or gear setting of instrument
- S_c = scale reading of instrument (a measure of existing torque)
- k = an instrumental constant derived from the rotor and cup dimensions

Table 3.--Linear regression coefficients for different levels of pattern coverage

Number of data sets	Coverage level	Density (*d*)	Brookfield viscosity (N_b)	Yield stress (*Y*)	Surface tension (*S*)	Modulus[1] of elasticity (*G*)	Regression constant	Regression coefficient
		g/ml	*Centipoise*	*Dyne/cm²*	*Dyne/cm*	*Dyne/cm²*		
	$D>1.8$	56.81	0.00446	0.0798	1.428		-142.0	[3]0.635
21	$D>t$ [2]	245.6	- .00446	- .00993	- .475		-158.2	[3] .758
	$D>t$	246.0	- .00539		- .519		-155.4	.758
	$D>1.8$	58.98	- .00190	.0444	1.152	-0.00188	-124.3	[3] .708
	$D>1.8$	55.44		.0233	1.450	- .00189	-142.2	.708
	$D>1.8$	70.66	- .00383	.0342	.986		-125.1	.554
	$D>1.8$	78.24	- .00775	.1099		- .00187	- 60.93	.699
	$D>1.8$	51.57	.00204		1.777	- .00187	-161.9	.705
	$D>1.8$		.00355	- .0165	2.662	- .00197	-172.2	.671
11	$D>t$	226.8	- .0295	.3512	-2.542	- .00375	9.186	[3] .929
	$D>t$	250.1	- .0334	.3310	-2.873		7.634	.802
	$D>t$	184.3	- .0166	.2066		- .00379	-130.7	.919
	$D>t$	168.1	.00168		2.412	- .00368	-288.4	.884
	$D>t$	171.7		.0237	2.108	- .00391	-270.1	.889
	$D>t$		- .00857	.1172	3.266	- .00411	-174.9	.813

[1]Modulus of elasticity (G) determined from measurements and calculations during subsequent tests and discussed later in this paper.

[2]t = trace

[3]indicates those sets of coefficients producing figures 8, 9, 10, and 11.

Table 4.--Rheological properties and pattern area data

Specimen code	Density (*d*)	Brookfield viscosity (N_b)	Yield stress (*Y*)	Surface tension (*S*)	Effective viscosity (N_e)	Modulus of elasticity (*G*)	Pattern area $D>1.8$	Pattern area $D>t$
	g/ml	*Centipoise*	*Dyne/cm²*	*Dyne/cm*	*Centipoise*	*Dyne/cm²*	-----*Percent*-----	
A.[1]								
Att-2A	1.041	490.0	37	71.6	20	648	11.7	58.8
CMC-1A	1.002	780.0	20	70.5	134	1,934	11.1	36.5
CMC-2A	.999	150.0	20	72.4	45	839	15.8	37.9
CMC-4A	.999	55.0	30	72.9	59	64	14.6	63.6
XA-1A	.987	9,050.0	800	71.0	137	14,970	5.8	10.0
XA-3A	.999	200.0	15	71.1	31	162	27.7	61.7
B.[2]								
PC-XA-K	1.071	800.0	63	71.0	125	649	20.0	65.0
PC-XA-K	1.074	1,500.0	100	71.0	191	819	22.0	60.0
PC-259	1.092	87.5	39	75.0	34	5,770	18.0	57.0
FT-100	1.105	1,600.0	193	77.0	78	1,523	37.0	77.0
FT-100	1.104	2,150.0	242	77.0	98	2,229	29.0	80.0

[1]The effective viscosity and modulus of elasticity were measured on replicate specimens during 1979. All other data are from table 2.

[2]All measurements were made in 1979, except that pattern area percentages are taken from table 2 using values (in some cases averages) from samples in table 2 of composition and properties most similar to those of the 1979 samples.

Recoverable shear (elastic strain, s) = ϕ / c

where: ϕ = angle of relaxation instrumentally observed (a measure of elasticity)

c = an instrument constant

Using these relationships the effective viscosity and modulus of elasticity are calculated:

Effective viscosity (N_e) = $N_a\ (1 + s)$

Modulus of elasticity $(G) = \frac{s\ a\ c}{\phi}$

where: a = an instrumental constant characteristic of the rotor and cup.

The second phase also included five commercial retardant samples, prepared and studied in 1979, in order to provide a wider range of density values. Density, Brookfield viscosity, and yield stress data for these were compared with those of similar materials in table 2. Estimated values of pattern area percentages were assigned to the 1979 samples (for which wind tunnel data could not be obtained) based on identical or averaged values of the pattern area from table 2. The complete data sets for these samples are shown in Part B of table 4.

The results of the regression analyses of the 11 data sets shown in table 4 are given in the latter part of table 3. Because the effective viscosity and the modulus of elasticity are both functions of the same two measured quantities, ϕ and S, the inclusion of both in a regression computation would be redundant. Values of the effective viscosity are given in table 4, but were not used in the computation; thus it does not appear in table 3.

Dispersal pattern area = D(density) + N_b (Brookfield viscosity) + Y(yield stress) + S(surface tension) + G(modulus of elasticity) + constant

Blanks in table 3 indicate those variables which were omitted for that computation. This allows some estimation of the relative importance of that variable to the quality of the correlation.

Figures 8, 9, 10, and 11 represent graphically the four computations indicated by footnote 3 in table 3. Each of the 11 (or 21) sets of data generates one point for each of the two pattern areas used.

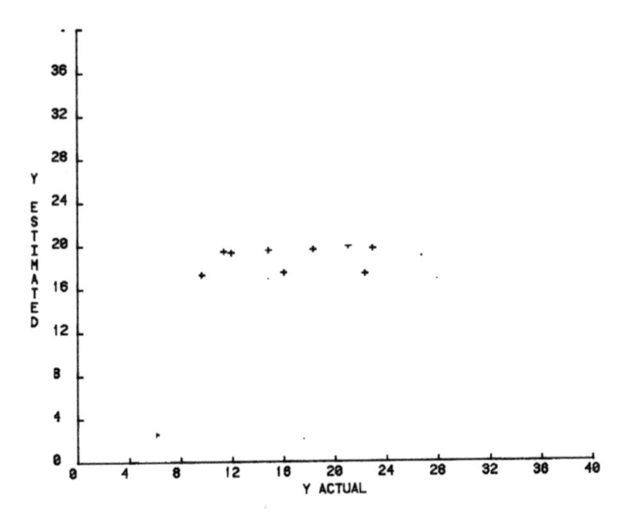

Figure 8.--Plot of pattern areas observed (actual) against those predicted by the model based on 21 data sets of four variables at density > 1.8 percent.

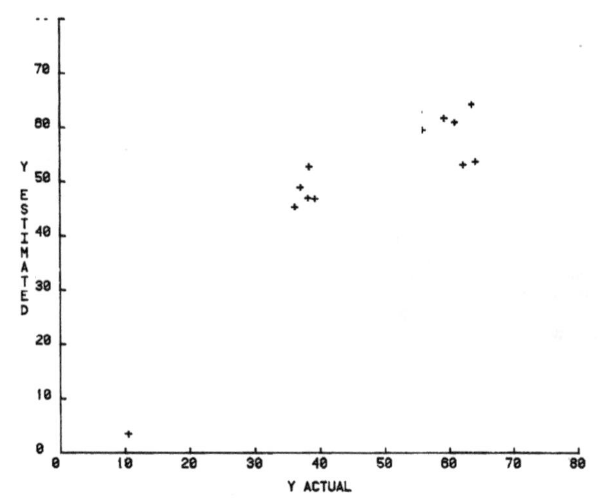

Figure 9.--Plot of pattern areas observed (actual) against those predicted by the model based on 21 data sets of four variables at density > trace.

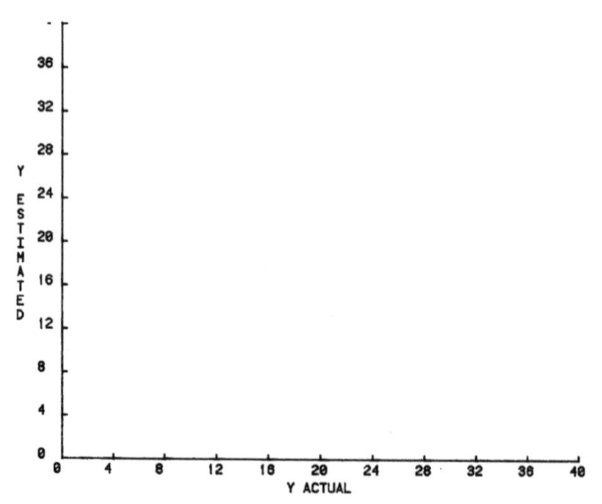

Figure 10.--Plot of pattern areas observed (actual) against those predicted by the model based on 11 data sets of six variables at density > 1.8 percent.

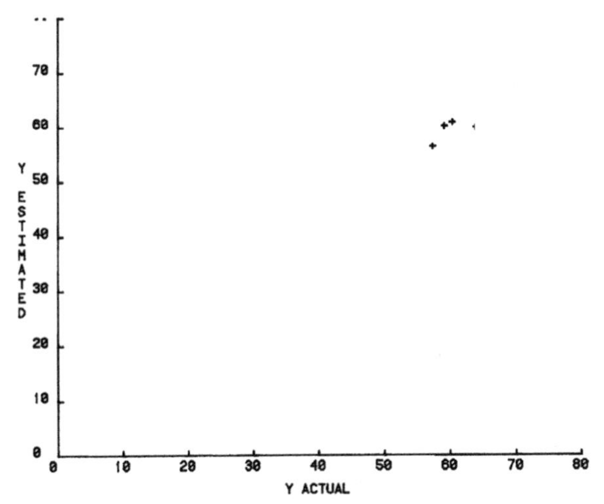

Figure 11.--Plot of pattern areas observed (actual) against those predicted by the model based on 11 data sets of six variables at density > trace.

CONCLUSIONS

The regression coefficients given in table 3 indicate that a definitely useful level of capability is present for predicting the relative size of drop pattern areas from physical and rheological data. The graphs of figures 8 through 11 (especially figures 10 and 11) make the same statement. The correlation is slightly better for coverage at the trace level.

The elimination of each of the variables, one at a time, indicates (table 3) that the modulus of elasticity is the variable to which the regression coefficient is the most sensitive. Density seems to be the next most influential variable. Deleting two or more variables leads to much more drastic decrease in the value of r^2.

Some retardant performance criteria can be modeled on theoretical considerations. An example is the treatment of droplet size by using fluid dynamic theory (Andersen and others 1976). The present attempt to relate pattern area to a diverse set of partially unrelated but conveniently available measurements has been approached in an empirical way, and the assumption of a linear relationship is arbitrary.

It is attractive to suspect that the intermediate coverage cited, 1.8 mg/cm², might correspond approximately to some level found to be effective in field practice. This is not correct, however; coverage levels seen in this study are five to 10 times lower than those recommended for actual wildfire control. It is equally important to note that the 30-inch (76.2 cm) drop distance in 40 mi/h (64.4 km/h) wind did not, at least for thickened materials, allow fully developed dispersal of the fluids into droplets of ultimate size. In most cases, the specimens were still being accelerated (sheared) by the horizontal airstream when they landed on the cup array.

It seems apparent from these results that a sizable collection of data from full scale airdrops should yield a model usable for maximizing airtanker performance. A significant reservoir of such data already exists, some as a result of studies directed toward the retardants themselves (George and Blakely 1973) and others toward the development of improved airtanker equipment and techniques (George 1975, Swanson and others 1978). Much of the necessary data is not contained in the reports themselves, but is on file at the Northern Forest Fire Laboratory (NFFL), Missoula, Mont. In using these data, rheological data would have to be obtained by preparing retardant specimens essentially identical to those used in the airdrops. This is feasible because stocks of identical component materials are in storage at the Northern Forest Fire Laboratory. In practical use, the model would need to take into account drop heights (altitude above terrain), airspeed, and details of the aircraft tank and gate system.

Publications Cited

Andersen, W. H., R. E. Brown, K. G. Kato, and N. A. Louie
1974a. Investigation of rheological properties of aerial-delivered fire retardant. Final report, Shock-Hydrodynamics, contract 26-3198 to Intermt. For. and Range Exp. Stn., 149 p.

Andersen, W. H., R. E. Brown, N. A. Louie, and others.
1974b. Investigation of rheological properties of aerial-delivered fire retardant, extended study. Final report, Shock-Hydrodynamics, contract 26-3198 to Intermt. For. and Range Exp. Stn., 66 p.

Andersen, W. H., R. E. Brown, N. A. Louie, and others.
1976. Correlation of rheological properties of liquid fire retardant with aerially delivered performance. Final report, Shock-Hydrodynamics, contract 26-3198 to Intermt. For. and Range Exp. Stn., 95 p.

Davis, James
1959. Air drop tests, Willows, Santa Ana, Ramona 1955-1959. 22 p. California Air Attack Coordinating Committee.

Garcia, John, and James D. Wilcox.
1961. Fall by canopy formation: a theoretical study. CRDL Special Publ. 1-30, ASTIA AD 271919.

George, C. W.
1975. Fire retardant ground distribution patterns from the CL-215 air tanker. USDA For. Serv. Res. Pap. INT-165, 67 p. Intermt. For. and Range Exp. Stn., Ogden, Utah.

George, C. W., and A. D. Blakely.
1973. An evaluation of the drop characteristics and ground distribution patterns of forest fire retardants. USDA For. Serv. Res. Pap. INT-134, 60 p. Intermt. For. and Range Exp. Stn., Ogden, Utah.

Hawkshaw, J. K.
1969. Advantages of the field membrane tank fire bomber system. Field Aviation Co. Ltd., Toronto International Airport.

MacPherson, J. I.
1967. Ground distribution contour measurements for five fire bombers currently used in Canada. Aeronautical Report LR-493, National Aeronautical Establishment, National Research Council of Canada.

Swanson, D. H., A. D. Luedecke, and T. N. Helvig.
1978. Experimental tank and gating system (ETAGS). Final report, Honeywell contract 26-3245 to Intermt. For. and Range. Exp. Stn., 284 p.

Van Wazer, J. R., J. W. Lyons, K. Y. Kim, and R. E. Colwell.
1963. Viscosity and flow measurement. Interscience, New York.

Wilcox, James, Joseph V. Pistritto, and Alan B. Palmer.
1961. Free-fall breakup of bulk liquids. CRDLR 3085, ASTIA AD 271920.

Van Meter, Wayne P., and Charles W. George.
1981. Correlating laboratory air drop data with retardant rheological properties. USDA For. Serv. Res. Pap. INT-278, 12 p. Intermt. For. and Range Exp. Stn., Ogden, Utah 84401.

Laboratory measurements of rheological properties of fire retardant materials have been correlated with wind tunnel area distribution experiments to develop a model that predicts the relative area of the pattern of retardant dropped from aircraft.

KEYWORDS: fire retardant, air drop, rheology

☆ U.S. GOVERNMENT PRINTING OFFICE: 1981-0-780-812

The Intermountain Station, headquartered in Ogd Utah, is one of eight regional experiment stations charged with providing scientific knowledge to help resource managers meet human needs and protect forest and range ecosystems.

The Intermountain Station includes the States of Montana, Idaho, Utah, Nevada, and western Wyoming. About 273 million acres, or 85 percent, of the land area in the Station territory are classified as forest and rangeland. These lands include grasslands, deserts, shrublands, alpine areas, and well-stocked forests. They supply fiber for forest industries; minerals for energy and industrial development; and water for domestic and industrial consumption. They also provide recreation opportunities for millions of visitors each year.

Field programs and research work units of the Station are maintained in:

Boise, Idaho

Bozeman, Montana (in cooperation with Montana State University)

Logan, Utah (in cooperation with Utah State University)

Missoula, Montana (in cooperation with the University of Montana)

Moscow, Idaho (in cooperation with the University of Idaho)

Provo, Utah (in cooperation with Brigham Young University)

Reno, Nevada (in cooperation with the University of Nevada)

correlatinglabor278vanm
ST000328

CPSIA information can be obtained
at www.ICGtesting.com
Printed in the USA
BVHW08s1101170918
527713BV00021B/599/P